Friedrich Uppenborn

History of the Transformer

Translated from the German.

Friedrich Uppenborn

History of the Transformer
Translated from the German.

ISBN/EAN: 9783337186050

Printed in Europe, USA, Canada, Australia, Japan

Cover: Foto ©berggeist007 / pixelio.de

More available books at **www.hansebooks.com**

HISTORY

OF

THE TRANSFORMER.

BY

F. UPPENBORN,

EDITOR OF THE "CENTRALBLATT FÜR ELECTROTECHNIK,"
AND CHIEF OF THE ELECTRO-TECHNICAL TESTING STATION IN MUNICH.

TRANSLATED FROM THE GERMAN.

E. & F. N. SPON, 125, STRAND, LONDON.

NEW YORK: 12, CORTLANDT STREET.

1889.

PREFACE.

As of late the employment of alternating current transformers has largely increased and become of great importance, indeed as they are called upon to play a striking part in electric lighting from central stations, the author has thought a short notice of the development of this invention would possess some interest. This task appeared to be so much the more pressing, as many distorted versions of the invention and its priority have found place in the technical journals.

The author has not let the reading of the large number of patents discourage him, and hopes that the following plain and concise statement of these researches will contribute towards the forming of a correct judgment as to the services rendered by the several inventors.

THE AUTHOR.

HISTORY OF THE TRANSFORMER.

As we wish to write of those discoveries which led up to the invention of the transformer, we must go back to a time, old as compared with the modern development of electrotechnics. For the starting-point of our observations we shall take Faraday, who, like Newton in mechanics, led the way in the domain of electricity, and whose name stands in the most intimate relations with all inventions for the mechanical production of the electric current, and therefore with the later development of electro-technics.

The most important discovery for which we have Faraday, 1831. to thank Faraday is that of induction. This discovery was made by him in the year 1831, and intimated to the philosophical world in a paper read on the 24th November, 1831, appearing in the Transactions of the Philosophical Society in the year 1832.

Faraday's first induction apparatus consisted of two coils of wire, the one being slid over the other. As he was passing the current from a battery through one of these, he made the discovery that each time the circuit of the coil was opened or closed an

B

electromotive force was created in the second coil, which caused a short gush of current or induction current to flow, provided the circuit of this coil was closed, as might be through a galvanometer. The peculiarity of this induced current was, that it only flowed in the second coil during the time the current in the first coil took to reach its normal strength after closing the circuit, or on breaking the circuit during the time the current took to decrease from its normal strength to zero.

This discovery undoubtedly belongs to the domain of the transformer, induction being the physical precedent upon which the transformer is based; indeed, a tranformer is in principle an induction apparatus.

Fig. 1.

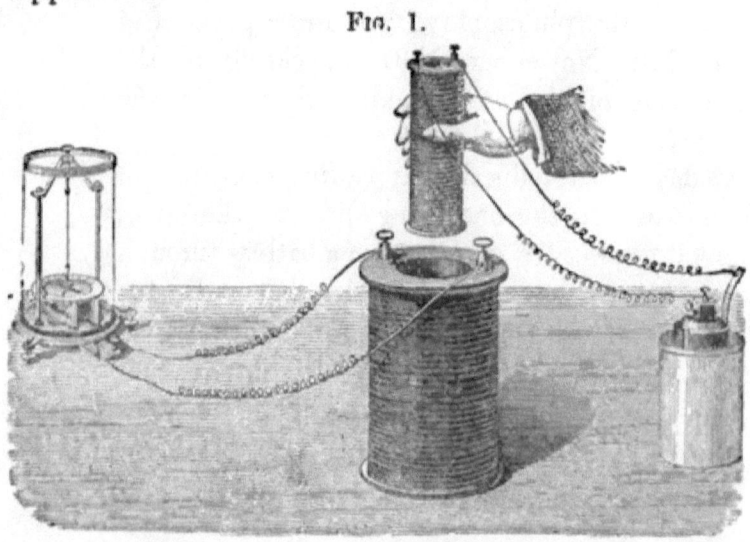

Fig. 1 represents the arrangement of this funda-mental experiment. The primary coil is connected with the battery, the secondary with the galvanometer. The primary coil, in order to obtain the best effect, is placed inside the secondary, and on opening and closing its circuit the needle of the galvanometer is thrown to the one or the other side respectively.

The arrangement, as in Fig. 2, made by Faraday showed itself to be an especially effective combina-

FIG. 2.

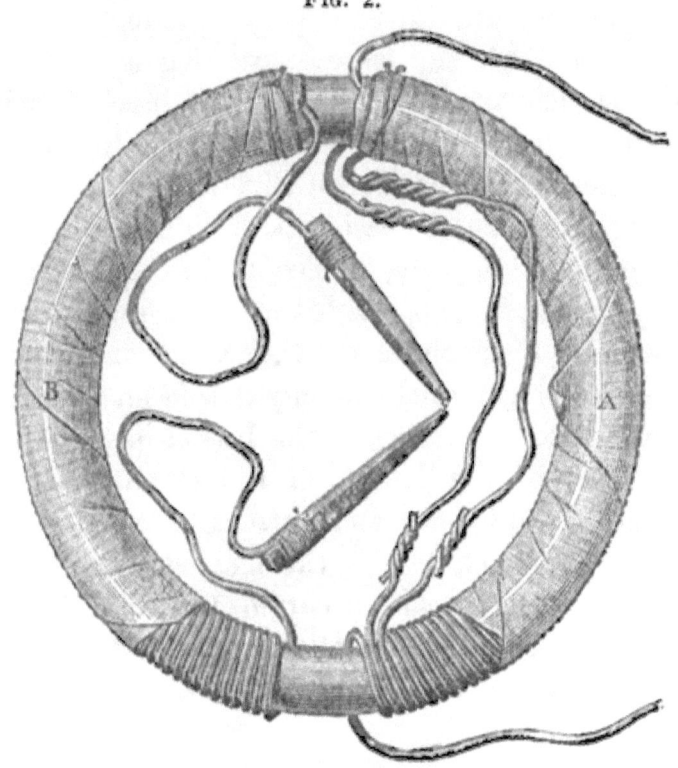

tion for the production of these induction phenomena. There were wound round an iron ring two separate wires of about the same length. The one coil was brought into connection with a battery, and to the

ends of the other a pair of electrodes were attached.
The current from the battery being sent through the
primary coil, lines of force were produced which ran
almost altogether in the iron core. As the core
possessed only a very small magnetic resistance, the
intensity of magnetisation was very great, and on
closing the primary circuit a strong inductive effect on
the secondary coil was produced. Faraday obtained
with this apparatus the first sparks of induction. The
apparatus is all the more interesting as, although
not completely without poles, it at least forms a
closed magnetic circuit. It has much likeness to
the non-polar transformer of Zipernowsky, Déri, and
Blathy, but it may be easily shown to be not entirely
poleless. Poles mean, in electrical as well as magnetic circuits, those points between which the greatest
difference of potential exists. A current without
difference of potential can only flow in an electrical
or magnetic circuit when the loss of potential in
each part of the length of the circuit, viz., the
product of resistance and current, is equal to the
gain of potential, that is the magneto- or electromotive force; therefore a current without difference
of potential requires that the resistance and magneto-
or electromotive force in each part of the length be
the same. Now the magnetic resistance of a symmetrical iron ring is constant in all parts of the
length of its magnetic circuit. In the case in
question only one half of the ring was excited,
therefore poles must have been formed at both ends
of the exciting coil. The ratio of transformation of

this apparatus of Faraday's was equal to unity, so it had therefore no claim to the designation of "transformer."

The induction apparatus of Faraday in its simplicity was in a certain measure the embryo out of which all dynamos and transformers have developed. We have seen how the first induction current was discovered by making and breaking the current from a battery in the primary coil. This method was at first adhered to, until Faraday remarked that when the secondary was quickly drawn out of or put into the primary coil, induced currents were also produced without requiring to break the circuit, the wires of the secondary coil cutting the lines of force in the magnetic field of the primary coil. He then replaced the primary coil and battery by a permanent magnet, which was likewise dipped into the induction coil, Fig. 3.

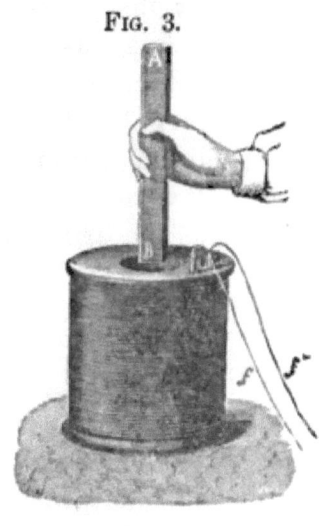

Fig. 3.

From this, and from the later development of this invention, it follows that the question was not of a transformer in the present sense of the word, but of a secondary generator. Transformers as at present understood were first known in Europe as the Ruhmkorff's induction coil. Before we take up this invention we shall mention a much earlier and like invention, which had already been made in the United States in the year 1838. This was the induction coil of Professor Page, and was the outcome of another invention by Professor Henry, whose apparatus was only a single induction coil. The

Henry and Page, 1836.

first public notice of Professor Page's apparatus appeared in the Silliman-Journal of 12th May, 1836, under the title, "Methods and trials of obtaining physiological phenomena and sparks from a heat engine by means of Professor Henry's apparatus." In May, 1837, Sturgeon published, in the "Annals of Electricity," in London, a description of the apparatus of Henry and Page.

Callan, 1837.

Callan, an English student of physics in Minnoth, showed first, in the year 1837, that if high tension was wanted, it was necessary to employ thick wire for the primary and thin for the secondary coil. Before this time wires indeed of different lengths, but of equal cross sections, had always been employed. His apparatus was not so bad as those before known, but still stood far behind that of Professor Page.

Page, 1838.

The arrangement of Professor Page's apparatus, which is shown in Fig. 4. was as follows :—Two coils of wire well insulated from one another were wound on to a bundle of iron wires. A self-acting contact-breaker was put into the primary circuit, and consisted of a double lever E, having on one of its arms two parts bent downwards, so as to dip into two mercury cups. The movement of the part H, as compared with that of E, was so small that it remained always in the mercury. At M, however, when the lever was set in motion contact was broken and made. To prevent oxidation Page poured in a layer of alcohol over the quicksilver.

FIG. 4.

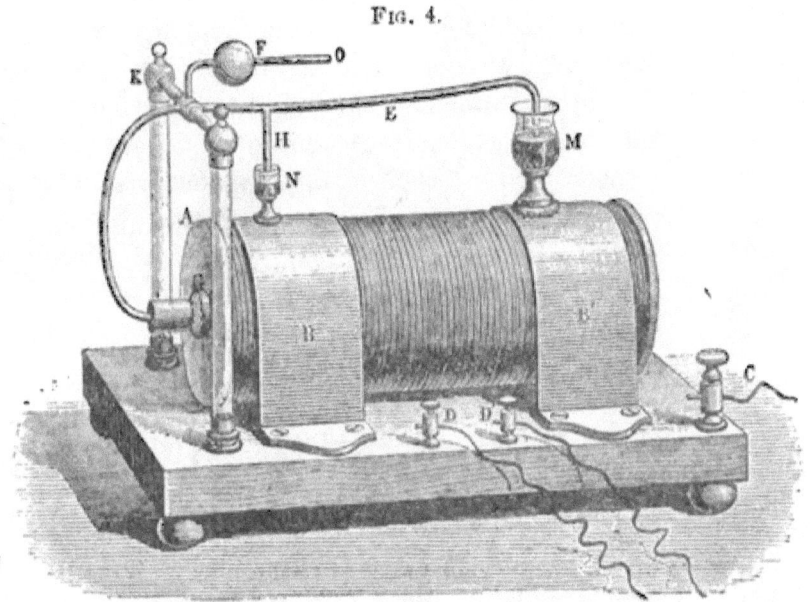

The continuation of the lever in the other direction of the axis, which was borne by two pillars K, was bent backwards, and on its end carried a cylindrical piece of iron standing before the end of the bundle of iron wires. If the primary coil were now placed in connection with a source of current, the iron core became magnetised, attracted the cylindrical piece of iron to itself, and by raising the lever E broke the contact at M. The iron core then lost its magnetism, released the iron armature, and the

play began anew. A counter-weight F, which could
be shifted along another lever O, allowed the play of
the contact-breaker to be regulated. It will be
found that this interruptor was very like that constructed many years afterwards by Léon Foucault.
The effects which Page produced by means of this
instrument were much more intense than those produced by Ruhmkorff with his, as Page succeeded
with only a single Grove element in inducing in the
secondary circuit such a high electromotive force as
produced sparks 4½ inches in length through a
vacuum tube—a result that Ruhmkorff, although
his invention created such a great and well-deserved
attention, did not attain. In the year 1850 Page
built a much larger apparatus.

In order to give some idea of the magnitude of the
electro-magnetic forces which came into play here,
suffice it to say, that the exciting coils could hold
suspended in the air in their interior an iron core
weighing 520 kg. The primary or magnetising coil
was of square copper wire, with a side measuring
¼ inch, and a battery of 50 to 100 Grove elements
was employed, the immersed area of the surface of
the plates being 100 square inches. This apparatus
gave sparks of great length. When, with maximum
curr rent strength, the primary circuit was broken,
sparks of 8 inch length were received.

Ruhmkorff,
1848.

Ruhmkorff constructed, in the year 1848, the so-
called spark-inducer named after him, the object of
which was also to convert currents of low tension into
hose of very high tension. With this coil and like

coils of larger dimensions effects were produced, but
only such as were afforded by the common forms of
frictional electrical machines. All things considered,
it is not a little surprising that while the invention of
the Rhumkorff coil was still in its infancy, the won-
derful output of Page's apparatus was still, even in
the year 1851, quite unknown in Europe.

Fig. 5 represents the earlier form of the Ruhmkorff
apparatus. It consisted of a bobbin of good insulating material; thoroughly dried wood, or better, hard

FIG. 5.

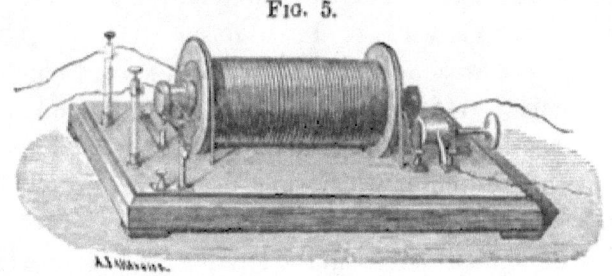

rubber. The two end pieces of the bobbin were usually
made of grooved glass discs, and were bound down to
the bedplate of the apparatus by two wires.   Inside
the coil was the already often-mentioned bundle of
iron wires. The primary or inducing wire was next
wound upon the bobbin. As this wire had to carry
currents of comparatively great strength, it consisted of only one or a few layers of thick wire.
The circuit of this coil was completed as far as two
terminals on the bedplate, first passing through an
interruptor like what has already been described.
Over the primary coil, and after a sufficient layer

of insulation had been added, the secondary wire
was wound. As this wire was destined for very small
currents, it was of as fine wire as it was possible to
wind. In order to obtain high potential it was
necessary that the secondary should possess many
turns. In the earlier coils a length of between
8 and 10 kilometres was used; in the coils now
made this length has been increased to between 50
and 70 kilometres. The ends of the secondary coil
were connected to terminals insulated on glass
pillars. It was not nearly sufficient insulation for
the secondary wire to be covered with silk, but
every layer was well soaked with dissolved shellac,
and then well dried as it should be. A condenser
in connection with the primary coil was placed under
or in the bedplate, which was usually a box. This
condenser was, and is still, often made thus:—On both
sides of a strip of paraffined paper, several metres
long and of convenient breadth, tinfoil is stuck, at
the same time leaving a sufficient margin of paper for
insulation. The whole is then folded together suitably.
The effect of the coil is substantially enhanced when
the sheets of tinfoil are each connected to the circuit
of the primary coil in such a way that the condenser
is in shunt to the interruptor.

In Fig. 6 is shown a newer form of Ruhmkorff's coil,
with an interruptor like the mercury contact-breaker
which we have before described. According as the
movable weight is raised or lowered, the oscillations
of the lever, and consequently the induced currents,
follow one another more slowly or more rapidly.

We find a further development or modification of C. T. & E. B.
the invention of Page and Ruhmkorff, patented by the Bright, 1855.
brothers C. T. and E. B. Bright on 21st October,
1852, and No. 2103 in the year 1855.   In the latter
of these patents the inventors state what follows concerning the nature of their inventions.

Fig. 6.

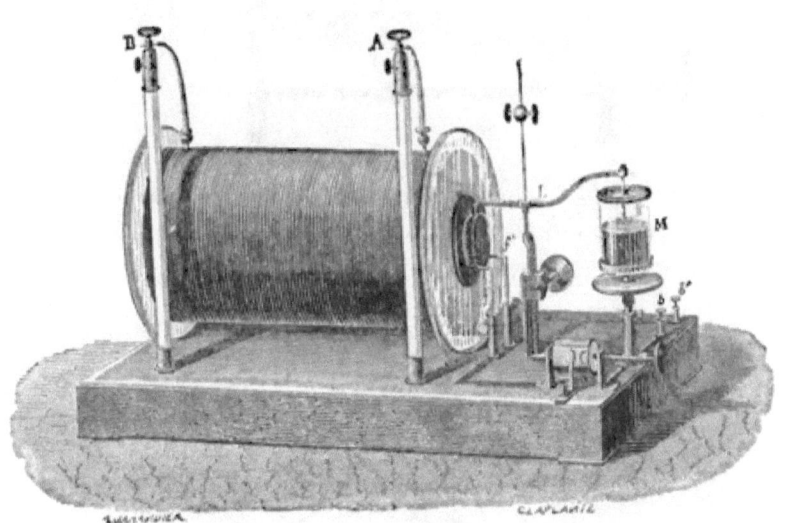

"A section of an induction coil made after this
manner is shown in fig. 7, having a very strong
effect.   The primary wire, of which only a part is
shown, is wound on an iron core, and outside is surrounded by an iron cylinder.   Both of these are
metallically connected by the flanges of the bobbin,
which also are of iron.   The secondary coil may also
be surrounded by an iron tube, and if the resistance
of the circuit be extraordinarily great with still more

primary coils, or it may be also contained in the
same tube as the primary. In cases where it is
found necessary to increase the quantity of the

FIG. 7.

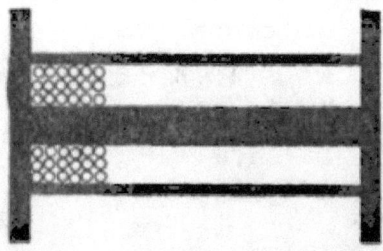

electro-magnetic effects, we find that the forms
shown in Figs. 8 and 9 are very effectual, and may
by varied on the same principle.

FIG. 8.

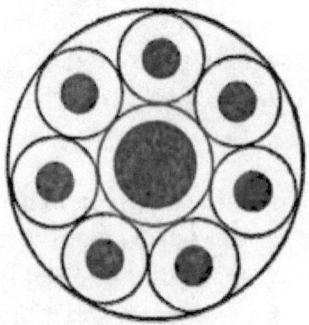

The iron core in the middle is wound with
the primary wire, and is surrounded by the other
iron cores, which are fixed into the large flanges of
the middle core, and carry the secondary coils.
Should still greater effects be required, more primary

or secondary coils connected in series with the others may be added outside, in order to produce a greater extension of the poles and a more extensive induction."

FIG. 9.

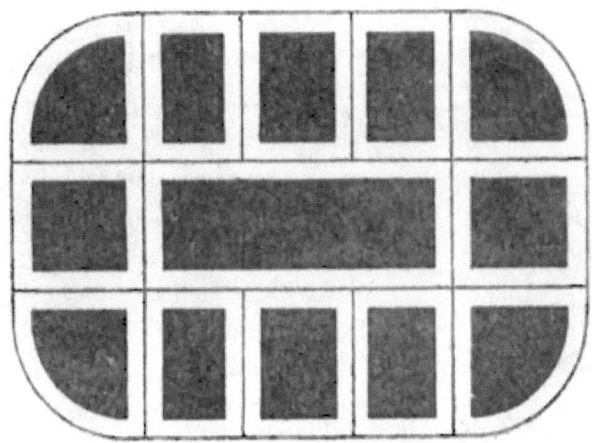

This patent is interesting also for the fact that in it we find a disposition of parts, viz. the arrangement of several induction coils in ranks, and connected with one another in parallel, which nearly 30 years later was taken up and practically used by Gaulard.

Among the patents of the year 1857 there is an English one by Harrison, claiming as its object the passing of a primary current through one or more induction coils, and the connection of the secondary coils with the carbons of an arc lamp. There is nothing remarkable in the description. *Harrison, 1857.*

The last attempt to use induction coils for industrial purposes is met with in the year 1878. In this year Jablochkoff took out a German patent, which *Jablochkoff, 1878.*

was also carried out in practice. He required currents of very high tension to feed his kaolin lamp; at that time such currents could only be produced by induction coils. He writes as follows in his patent:—

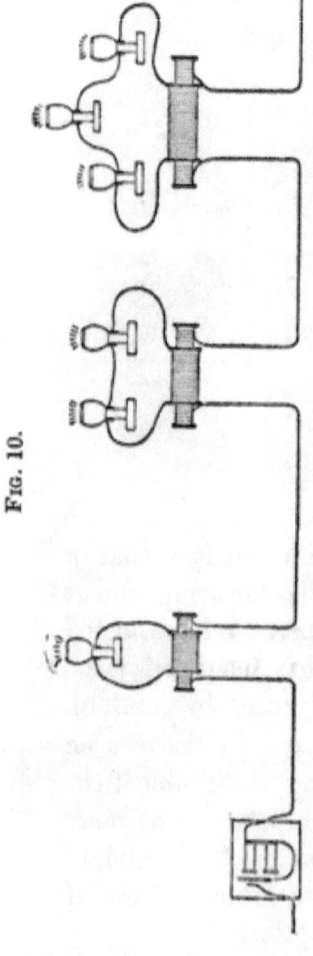

Fig. 10.

"Die Herstellung einer elektrischen Beleuchtung nach meinen System begreift eine Serie von Induktionsrollen in sich, wovon die inneren Drähte in eine elektrische Leitung eingeschaltet sind."

Jablochkoff used intermittent direct currents as well as alternating currents. The arrangement shown in Fig. 10 was for the former. He states concerning this:—

"In diesem Falle sind die Induktionsrollen mit Unterbrecher und Kondensator ausgestattet, oder man kann auch, wie die Zeichnung nachweist, einen und denselben Unterbrecher für alle Rollen anwenden. Die Induktionsrollen B¹ B² B³, nach einen beliebigen Prinzipe konstruirt, sind in der Nähe der Lichtherde angebracht."

Concerning the employment of alternating currents, Jablochkoff says :—

"Diese Disposition weicht von der ersteren nur
durch die Weglassung des Unterbrechers und des
Kondensators der Rolle ab.

FIG. 11.

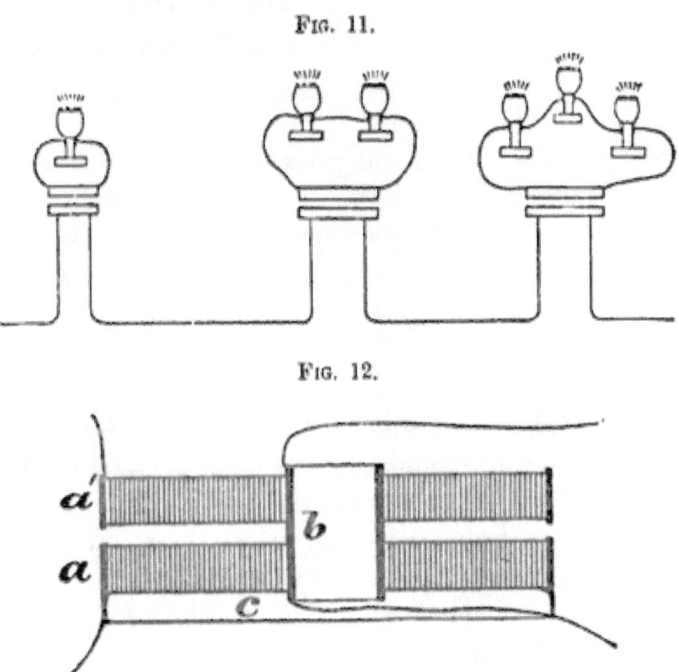

FIG. 12.

"Die in Fig. 11 angewendeten Rollen sind in
Fig. 12 detailliert gezeichnet. Auf einer kreis-
förmigen Scheibe C aus weichem Eisen erhebt sich
in der Mitte derselben ein hohler Cylinder b aus
Holz oder anderem isolirten Materiale; um den
unteren Teil des letzteren ist die Hauptspirale a
gewickelt, welche aus bandförmigen Kupferstreifen

oder anderem Metalle besteht. *a'* ist die in gleicher Weise zusammengesetzte Induktionsspirale, deren Drahtenden zu den Lichtherden führen. Zwischen den einzelnen Windungen der Spirale sind Streifen aus Papierkarton oder einem anderen isolirenden Material angebracht. Die Spirale *a* ist in die Hauptleitung, wie Fig. 11 zeigt, eingeschaltet. "

The second claim of this patent is also interesting, and reads as follows:—

"Die Einführung einer Serie von Induktionsrollen in den Umkreis eines beliebigen Elektricitätsgenerators zur Erzeugung einer Serie von Induktionsströmen, welche es gestatten, Lichtherde von verschiedener Intensität durch eine einzige Elektricitätsquelle zu versorgen, was zur vollständigen Teilbarkeit des elektrischen Lichtes führt."

Jablochkoff's system as just described was to be seen working in the Paris Exhibition of the year 1878. A proper industrial application of this system does not appear to have taken place.

C. T. & E. B. Bright, 1878.

In the year 1878 the brothers Bright had also made further progress in the use of induction coils for electric lighting purposes, and in the same year they took out the English patent No. 4212, in which they described the use of alternating currents for working secondary apparatus or induction coils placed at various points where light was required. We shall here quote some very interesting sentences from this patent, which again show that the brothers Bright knew already in the year 1878 the properties of transformers suiting them for electric lighting purposes; indeed they then anticipated the principles contained in the later patent of Gaulard. Here is an abstract from the description :—

"At each point where electric light is used, the electric lamps or groups of such lamps are fed by the secondary coil or coils of an induction apparatus placed there. The primary coils of all the induction apparatus are in the common circuit of one main-lead, which is in connection with a battery or a magneto-electric machine placed in some suitable situation. The size and length of the primary and secondary coils of each induction apparatus is determined according to the number of lamps at each point, where the secondary current shall supply the electric lighting."

The employment of induction coils for the distribution of light, heat, and power was patented in England in the same year by Edmund Edwards and Alphonse Normandy. Among other matter in this patent there is as follows :— E. Edwards & A. Normandy, 1878.

"At or near every point where it is required that a light shall be produced, we arrange a coil (or series of coils) of insulated metallic wire or ribbon (preferably surrounding a bar or wires of soft iron), through which coil or coils the current from the principal wire first described can be passed when desired, or cut off by means of a key or lever. Round, or adjacent to, each coil of insulated wire described, we form one or more secondary coils of insulated metallic wire, or ribbon, arranged so that the passage of the rapidly intermittent current of electricity, as described, through the primary coil or coils, generates a corresponding current of electricity in each of the secondary coils.

Strumbo, 1878.

In the same year, Strumbo had also constructed a secondary generator like that of Gaulard, and a description of it was contained in the newspaper 'Le Monde,' of 24th October, 1878. It is of note in this apparatus, which we have illustrated in Fig. 13, that

FIG. 13.

the primary and secondary wires were wound side by side, and that both coils had the same relative position to the iron core.

Harrison, 1878.

Harrison also, in the same year, took out a patent having the same object as his of the year 1857. Both patents proposed the connection of induction coils in series. This is especially clearly mentioned in the latter of these, as there he says that both induction coils are put in circuit at intervals along the main-lead, or primary circuit, so that one or more coils are near the places where lamps are to be fed.

We find in Meritens' English patent, No. 5257, of the year 1878, the series connection of primary coils in the dynamo-circuit also described.

Meritens intended to employ, in place of the many separately insulated circuits of the alternating dynamos of that time, only a single circuit, fed from one large or several smaller dynamos. A large number of induction coils connected in series, were to have been distributed in the different districts of a city. Besides this, Meritens made a combination of the secondary coils, so that he was in a position to produce currents and potentials of various dimensions.

We now come to an inventor, who, in his time, exercised a great influence upon electric lighting by means of transformers, and whose system was in every way a great advance on those of his foregoers. This man was named Jim Billings Fuller. He began to study electric lighting in his laboratory at Brooklyn in the year 1874, giving his whole energy for this object. Fuller's system of current distribution was first patented in America in the year 1878. The patent No. is 210,317, of 26th November, 1878. His apparatus is represented in Fig. 14. It consisted of an induction coil on which an electric lamp was mounted, to all appearances a Jablochkoff candle. The induction coil, to which we shall return later on, was built in the form of two horseshoe magnets joined together, and having consequent poles at the small coils in the middle, after the manner of the magnets of a Gramme machine.   The four large coils are the primary or exciting, the four small coils on the poles of the double magnets are the secondary coils.

FIG. 14.

The lever MN was of iron, and served to weaken the effects of induction, inasmuch as it formed a magnetic short circuit.   Here we find for the first time the employment of a regulating device. Fig. 15 illustrates the method of connection.*   As already mentioned, Fuller succeeded in setting aside many of the defects which were adhered to in the many very badly constructed transformers of his predecessors.   While he was busy carrying his invention into practice, he became a sacrifice to his over-great activity, and on the 15th February, 1879, he was taken away by illness.   Only a few hours before his death, he called his foreman to himself, and explained to him the principles of his system.   After ending his explanations, he asked him if he had understood all that he had said, and, on receiving an answer from him that he had, he smiled con-

* See also 'Scientific American,' 5th April, 1879, p. 212.

tentedly, and a few moments later he ended a useful life, which had given so much promise of good results.

In the year 1880 Edward Henry Gordon took out E. H. Gordon, the English patent No. 41,826. Gordon had con- 1880. structed an electric lamp based on the fact that when

FIG 15.

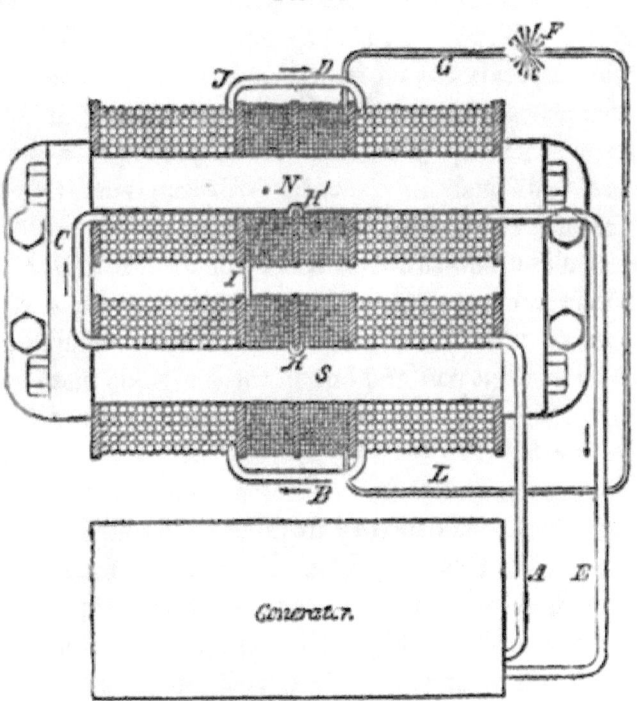

a current of sufficient electromotive force was passed over the space between two balls of platinum or platinum iridium, the balls were rendered glowing white. These balls were suspended by thin platinum wire, or the supports were of platinum, serving also to carry the current. For the production of overspringing sparks, it is well known that a great difference of potential is necessary, so Gordon was obliged to have recourse to induction coils, which he intended to excite by means of magneto-electric machines, or alternating current dynamos. In his patent he describes how this idea should be carried out, and he actually did feed two lamps of 50 c.p., or one of 100 c.p. The apparatus is described as follows:—"The primary consists of a bundle of iron wire 1·3 inch diameter, and 18 inches long. Three layers of insulated wire 0·08 inch in diameter are wound on it. The secondary is wound on an insulating tube, and consists of about two-thirds of a mile of wire 0·0075 inch diameter, covered four times with silk. It is wound in 60 discs." "There are three binding screws, one at each end and one in the centre, so that the whole coil, or either half separately, can be used for one lamp."

We do not find in Gordon's patent the slightest indication which would justify us in ascribing to him the invention of a system of distribution by transformers as known at the present day, but, on the contrary, it is clearly shown that the fundamental conditions of such a system of distribution were unknown to him, for he laid the chief weight upon connecting the induction coils in series, and on the production of high electromotive force necessary for his lamp. Over and above this, he was of opinion, as he stated prominently, that the more advantageous kind of dynamo was one such as that of de Meritens, having many coils of thin wire, which were connected to separately insulated leads.

Let us look back upon the inventions which were made in the domain of electric lighting by transformers from the time of Faraday's discovery of induction up to the year 1880. There we see that three distinct characteristics were possessed by all the systems invented up to that year. These three characteristics lay in the construction, the ratio of transformation, and the method of employing the transformers. Single transformers, with two or more poles, were used. The ratio was either 1:1, in which case the induction coil is really not a transformer, or it was from a low to a high electromotive force; but nowhere do we find that currents of high electromotive force were converted into those of low electromotive force. The idea in the use of transformers was that of division, not that of distribution of electric energy. The difference between division and distribution of electrical energy is, in the main, as follows. By a division of electrical energy it is meant that a fixed amount of produced energy is divided into pre-determined parts of a certain number and size, while it remains indifferent, as far as the total energy is concerned, in what manner and how many of these parts are usefully employed. By distribution of electrical energy it is meant, on the other hand, that the energy produced is variable according to the variable requirements of consumption, the maximum requirement being predetermined from the number and size of the local requirements, which also vary relatively to one another. Of the last of these systems there is no indication in any of the inventions of induction coils up to this date.

If we seek for the cause of these characteristics, we find that the reason why transformers with two or more poles were constructed is, that the electricians of those days either did not know or did not understand the principles on which a proper transformer should be constructed. With them the idea of a magnetic pole acting on a wire near it was always present, while they entirely overlooked the fact that the electro-magnetic force, not the pole, produced the electromotive force in the wire. On account of this they were of opinion that free poles in a transformer were not only not a drawback, but, on the other hand, a distinct advantage.

We find that Fuller especially held this view. He sought not only to have in his apparatus two simple poles, but double poles, and indeed he patented this arrangement of his transformer. The first claim of his patent reads thus:—

" The double electro-magnet herein described, the main coils of which are included in the circuit of a main conductor from a generator of alternating electric currents, producing in said magnet consequent magnetic poles, as shown, and around which poles are coiled helices of wire for receiving the currents induced by the polar changes, said helices being included in the local circuit with the lamp."

We must bear in mind that, as far as the ratio and

idea of employment of a transformer are concerned,
the problem at that time was quite another to what
it is now. At present the transformer serves princi-
pally to render possible the carrying of the current
to a great distance economically. The electricians
of those days were not so far advanced as to be able
to run arc lamps independently of one another on the
same circuit, and this they held to be quite impossible, whether the lamps were connected in parallel
or series. That apparatus was thought to be good
which allowed separately insulated currents to be
led from one source of current, each separate circuit
going to feed a single lamp. The chief reason for
this view lay in the fact that the extinguishing of
all the lamps in one circuit could easily take place
through the fault of one of them. At that time,
when an arc lamp was cut out of circuit, it was
replaced by a fixed resistance, instead of which it
was thought that induction coils would have suited
well. It may be casually mentioned that owing
to this fact too sanguine hopes of the solution
of the problem of independent working of lamps
were aroused, through a want of sufficient know-
ledge of the laws of induction. There have also
been apparatus other than induction coils used for
the purpose of making the points of consumption
independent of one another. We can only now
recall the patent of Jablochkoff, No. 1638, which
is based on the principle of connecting condensers
into branches of a quickly alternating main current,
from which arc lamps, &c., were fed; also a like

arrangement by Avernarius (Figs. 16 and 17), with
the use of secondary batteries, which were to be employed for either parallel or series connection.[*]

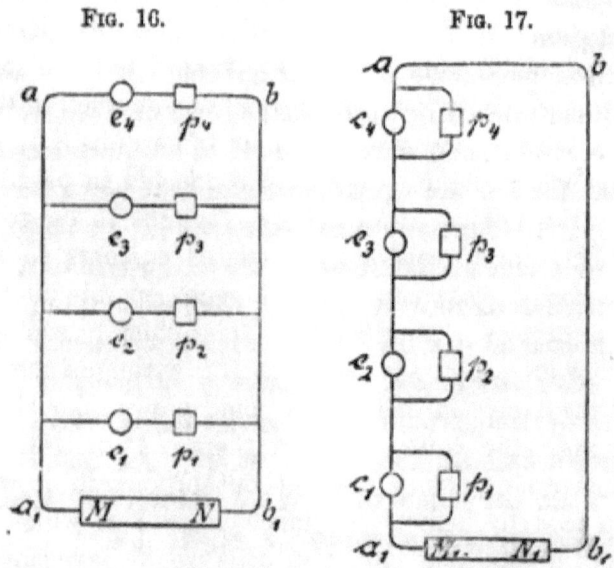

Fig. 16.   Fig. 17.

There were no transformers in those days which,
in the present sense of the word "transformer," convert high electromotive force to low to suit the
consumers. On the contrary the apparatus, which
was then used in electric lighting plant, was such as
converted low into high electromotive force, or such
that the ratio was 1 : 1, or nearly so, according as it
was determined by the connection in series of the
primary coils, and the difference of potential at the
consumption devices; for example, the induction
coils of B. Ruhmkorff, Jablochkoff, and Gordon.

[*] Avernarius, Centralblatt für Elektrotechnik, vol. iii. p. 323.

When, however, the term high electromotive force is met with in descriptions of the apparatus of that time, it must be taken to mean a great difference of potential between the terminals of the dynamo, not between the primary terminals of the transformers. Take, for instance, 100 transformers connected in series, run with a difference of potential at the dynamo of 1000 volts, although it was not known at that time how to produce so high an electromotive force, still this would give across the primary terminals of each transformer the modest difference of potential of 10 volts. In this way the difference of potential at the generator was determined by the number of transformers in series. This system had plainly the great disadvantage, that no matter how tortuous a path the lead must follow, it had to pass through the primary coils of all the transformers, and the principles of a proper system of distribution were not present.

With the invention of the incandescent lamp the activity of inventors was given quite another direction. The systems of electric lighting up to this time were not sufficiently advanced to permit even of a *division** of the electric light, that is, the ability to feed even a small number of lamps from one generater. We shall only mention this invention so far as it helps to further the history of the transformer.

Gramme made the earliest arc lamp that could be employed alone; then followed Jablochkoff, as the

* At that time a customary and very characteristic expression.

first who carried out practically, and with good results, the use of arc lamps in series or in parallel arc with condensers. Siemens and Halske then replaced the Jablochkoff candles with their differential lamp, which, although not offering an opportunity for a good division of light, was unexcelled in construction and manufacture, pointing out the way for further progress in arc lighting. This class of lighting was brought nearly as far forward as it is to-day by the introduction of continuous currents for this use by Brush.

With the invention of the glow lamp quite other aims were placed in the foreground for the electrical world. The incandescent lamp did not possess that unsteadiness of light which, with arc lamps, gave so much trouble to electricians. The prominent qualities of the glow lamp offered opportunity for the solution of a problem, such as gas had already solved half-a-century earlier, namely, the distribution of the electric light, or, more properly, of the electric current. For this, the already known and generally employed methods of connection were no longer sufficient. Edison was the first who demonstrated that the series method of connection was not suitable for glow lamps; at the same time he showed the advantages of parallel connection, coming forward with a thoroughly well thought and worked out system of distribution. By this means the change was made, and, from this time onward, all inventors were obliged to suit their systems to the demand, that each point of consumption must remain undisturbed by the variations of current which take place in the circuit.

Marcell Depréz has laid down in a work of his,[*] the laws which make it possible to hold the points of consumption of electric energy independent of one another, and, excepting some inexactnesses which crept into his representation, these laws have been almost all carried out in practice since that time.

The system of direct distribution to glow lamps had the one well-known serious drawback, viz. that it only allowed of limited employment, because the cost of the leads, with equal loss of energy, increased with the square of the distance from the source of current.

It was therefore obligatory, in order to carry the current economically to greater distances, to seek new means and ways, without rendering inefficient the only practical system of connecting incandescent lamps in parallel. The experience which had already been gained in the economical carrying of high tension currents with arc lamps in series, pointed out that high tension currents should be used, and that in the secondary circuits of transformers fed by such a current, consuming devices could be connected as might be desired.

Haitzema Enuma, in the year 1881, was the first H. Enuma, to go in this direction, and took out a patent for the [1881.] feeding of glow lamps by means of transformers.

He followed the principle of making each secondary circuit and each point of consumption independent. The means to this purpose which he thought to employ were not practical, and did not at

* Comptes Rendues, 1881, p. 872.

all differ in substance from those of his predecessors. His system is remarkable for his method of connecting the induction coils in the main lead, i.e. in series, using the secondary currents from these coils to excite other coils from which tertiary currents were received, and these again were further used to excite quaternary currents, and so on. This procedure stands on a level with that of the famed dynamo-electric chain of Siemens and Halske, of which it has been asked, " To what purpose ? "

The peculiarities of the system of Haitzema Enuma become evident from the following extract from his patent :—

" Solche (nämlich die bekaunten) Induktionsrollen werden in den Hauptstromkreis überall eingeschaltet, wo der Strom abgezweigt (!) werden soll; und durch diese Einrichtung erhält zuletzt jede elektrische Lampe, oder jeder durch Elektrizität in Betrieb gesetzte Apparat seinen eigenen Strom."

Haitzema Enuma had intended, so far as this shows, to connect the primary, secondary, tertiary, &c., coils in series, and the main lead being a closed circuit, the ends were taken to earth. The ends of the circuits of the secondary, tertiary, and further induced currents, were also connected together, or to earth.

Gaulard and Gibbs, 1883.   The first who came forward with an industrial employment of the series system were Gaulard and Gibbs, who, in the year 1883, placed before the public an installation of electric lighting in the Royal Aquarium in London.

There were two such apparatus as shown in Fig. 18,

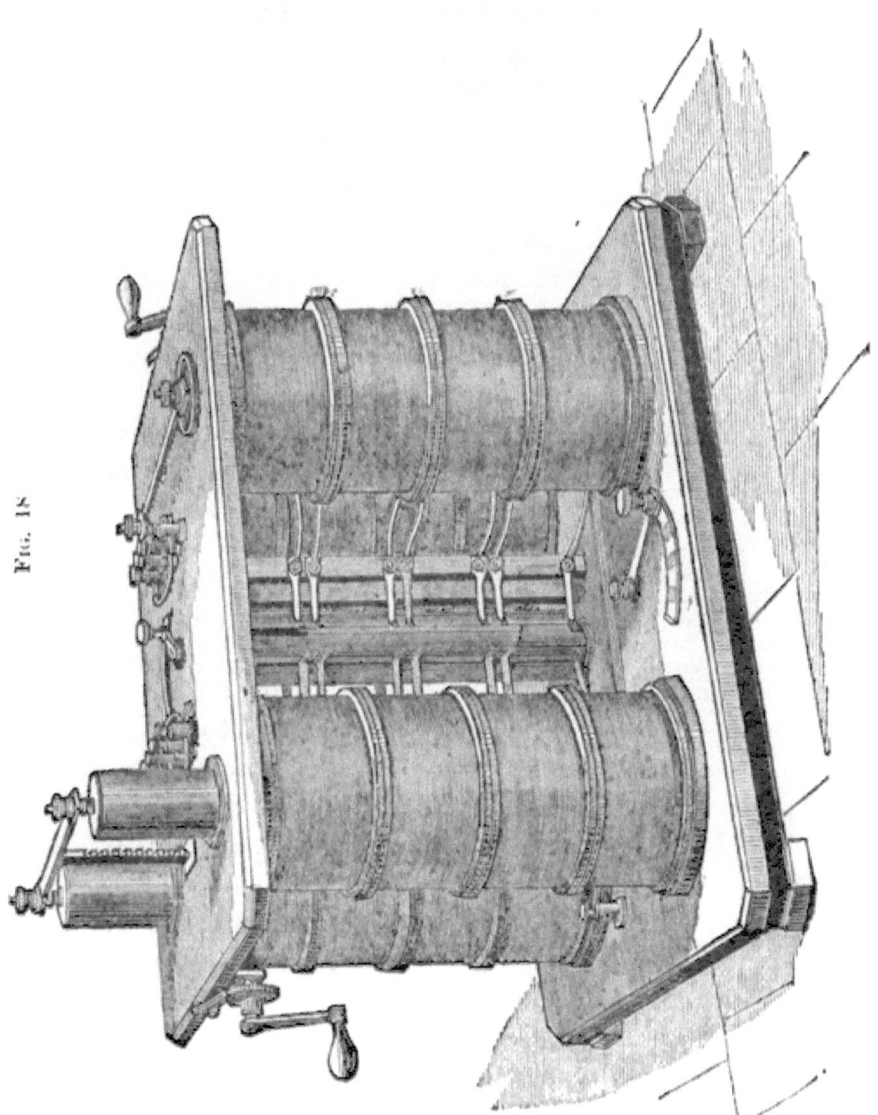

which were connected in series, and excited with
13 ampères from a Siemens' alternating current
dynamo. The apparatus had the following construction:—The induction coils, a section of one of which
is shown in Fig. 19, had three layers of primary wire,
and the secondary was wound in four divisions, the
ends of the wires of the divisions being led to a commutator. Fig. 20 shows this commutator placed in
the middle of four induction
coils. The ends of the secondary wires were connected to
eight terminals on the upper
plate of the apparatus, from
which the current could be
led away from each pair, or
combined at will. By aid of
the commutator, the number of
coils in circuit could be altered
as desired. On the lower
plate there was a second commutator, which served the

FIG. 19.

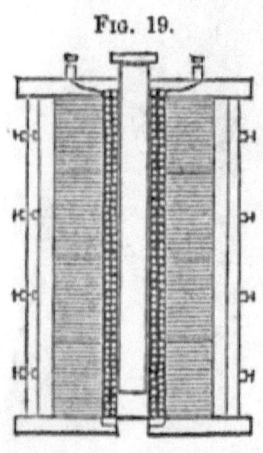

same purpose for the primary circuit.

The core of the apparatus consisted of bars of
insulated iron, and by means of a rack could be
raised or lowered in the coils for the regulation of
the current. Both of these arrangements had been
already long known.

In the same year another installation for the
lighting of some stations on the Metropolitan Rail-
way was taken in hand and carried out.

The source of current was a Siemens' alternating

FIG 20.

current dynamo of type *Wo*, which was excited by a
continuous current machine.   The potential was
supposed to be 1500 volts and the current 11·3
ampères. The main lead connecting the transformers
in series was of 7 wires of 1·5 mm. diameter, and
was 22·9 kilometres long, having a resistance of
30 ohms.   Three stations were supplied.   At Edge-
ware Road twelve coils, with their secondary coils in
parallel, fed 30 glow lamps; and other four coils, also
in parallel, fed two Jablochkoff candles. In Aldgate
two coils supplied one arc lamp, and twelve more
coils 35 glow lamps, each of 20 c.p. and three of
40 c.p.   At Notting Hill there were 22 glow lamps
and one arc lamp.   In this last installation coils
were employed with their coils aranged after a
somewhat different manner. On a pasteboard or
wooden cylinder of about 50 cm. in height a cable
was coiled in layers.

The interior of this cable consisted of a 4 mm.
copper wire well insulated with paraffined cotton,

Fig. 21.   and around this, parallel to its axis, lay
6 cables or cords, each consisting of
12 wires, also insulated with paraffined
cotton (Fig. 21). The wire of 4 mm.
formed the inductor through which
the primary current was passed.   The
six cables, each of twelve strands, formed the induced
portion of the apparatus, and the ends were connected to a commutator, so that they could be used
either in parallel or series.

The methods of construction and connection used

in these attempts by Gaulard and Gibbs did not differ in principle from those of their predecessors. Gaulard and Gibbs also employed in these trials bi-polar induction apparatus. The efficiency of such apparatus can only be comparatively small, because the effects of magnetisation, and therefore of induction, are weakened to a great extent by the lines of force having to pass for the greatest part of their path through air instead of iron. Taking another view of such apparatus, as they have a ratio of transformation of 1 : 1, they must, with the employment of high potential, be connected in series.

Undoubtedly Messrs. Gaulard and Gibbs have in their time claimed certain things as new and of their own invention, namely, the arrangement of several separate induction coils together, the placing of the coils next to one another, and the winding of the wires parallel. These claims, however, have been condemned from all sides as unjustified. The employment of several coils has already been mentioned as patented by the brothers Bright on 21st October, 1852, and was again later on discovered by Poggendorf, Ruhmkorff, Foucault, and others. We have also shown, on page 11, that the placing of the coils next one another had likewise been invented by the same men 30 years earlier. The symmetrical arrangement of both coils, the primary and secondary, had also already been used. (See page 18.)

But when, in spite of all this, we find Mr. J. K. Mackenzie* maintaining that the Fuller transformer

* The 'Electrical Engineer,' 17th Feb., 1888.

was non-polar, and further, that the following improvements must be ascribed to Messrs. Gaulard and Gibbs, viz. :—

1. The reduction of the primary and secondary wire-resistance to a minimum.

2. The attainment of the greatest possible coefficient of induction with the lightest apparatus.

3. The symmetrical arrangement of both coils.

4. The proportioning of the coils, so that the weight of metal in each is the same.

Seeing this, it must be thought that this gentleman either does or will not, understand the subject. Then if Gaulard has succeeded with his apparatus in obtaining some advantages as proposed in the above-mentioned clauses, Nos. 1 and 2, these advantages can be obtained to a much higher degree with non-polar transformers. This has been proven by Prof. Ferraris.*

The improvements mentioned under Nos. 3 and 4 are only to be attained with bi-polar transformers after difficult and otherwise disadvantageous arrangements; for instance, the combination of the primary and secondary wires in a common cable, or, when the coils consist of ribbon wire, by the winding of the one inside the other. With non-polar transformers these improvements are already inherent. The Fuller transformer was just as much without poles as two horseshoe magnets are, with their like poles laid together.

In all these systems with series connection of the

* La 'Lumière électrique," fol. xvii., p. 145-148, 1885.

transformers, the intensity of the current in the primary circuit must be held constant in order that it may be possible for the induction apparatus to maintain the secondary electromotive force constant. Notwithstanding this, constancy was not attained, but only one cause of the variations annulled. Another cause of the variations of the difference of potential at the secondary terminals of the coil still remained; this was the loss of potential due to resistance and self-induction, which increased with the load. The electromotive force of the secondary, and therefore of the primary coils, accordingly increases as the current in the secondary decreases. When no secondary current is flowing, the electro-motive force in the primary and secondary coils is a maximum. We have consequently this disproportion that the smaller the output of the apparatus the greater the energy consumed. With the secondary circuit open and a constant exciting current, the energy used could be as much as ten times as great as under full load.

The disadvantages of this system are apparent; for, putting aside the loss of energy arising from the disproportion between produced and consumed energy, each change of load on the secondary circuit exerted a great influence on the primary circuit, and again on the secondary circuits of the other coils in the main circuit.

All the transformer systems already described were intended, as we see, for subdividing the current, and as fitting therefor we find the series method of

connection universally brought forward. With this
method, owing to a rise in electromotive force which
was dangerous to the lamps, &c., when only a part
of those in the secondary circuit were extinguished,
it was compulsory either to run the induction coil
fully loaded or quite empty. Thus, when the number of lamps or other devices in use was varied, a
regulation of the current strength and uniform
working was either quite impossible, or only partly
possible by unreliable and incomplete mechanical
means. On this account no one succeeded with this
method in carrying out a rational distribution of
current by means of induction coils such as are required by the widespread demands for electric
current from a central station.

The first to point out the disadvantages of the
series method of connection was Rankine Kennedy,
who had devoted himself wholly to the study of induction apparatus. These disadvantages he published
in an article in the " Electrical Review " of 9th June,
1883. At the end of this article we find the interesting statement that transformers, when not
connected in the primary circuit in series, as had
been usual till then, but in parallel, form a self-regulating system of current distribution. Rankine
Kennedy expresses this in the following words:—
" In parallel arc, however, the secondary generator
is a beautiful self-governing system of distribution."
At the same time, however, his article affords proof
that the author then possessed only a limited com-
prehension of the physical facts concerned, because

he maintained, for instance, that the introduction of an induced counter electromotive-force in the circuit of an alternating current dynamo might constitute a means of regulation without loss of energy ; however, it might be allowed, that he meant by these words one of these elements which must be present in a really rational system of distribution with the use of transformers, if it were not the case that at that time he was not aware both of the properties of trans- formers suiting them for such a connection as well as those which make them self-regulating in a system of distribution. Above all this he had at that time never thought of a transformer in the sense, the word is used to-day, that is, as an induction apparatus, which converts high into low tension currents. This is quite clear, as is seen from the end of the sentence before cited, as he says, " But what about the size of conductors for such a system? Prodigious!" Kennedy thought to all appearance that the parallel connec- tion of transformers made possible self-regulation in the same manner as the simple direct parallel con- nection of incandescent lamps. While at the same time he imagined that on account of the small re- sistance of each coil the resistance of the net of leads must nearly vanish, therefore he concluded that the parallel connection of such induction apparatus as he had in his mind's eye was impracticable.

The apprehension of Kennedy's ideas, as we have here stated, finds direct confirmation from the leading article in the " Electrical Review " of 9th June, 1883. At the end of this leader the editors say, that " Mr.

Kennedy's apparatus is an induction coil pure and
simple." "Messrs. Gaulard and Gibbs will scarcely
deny, nor can they deny, that the action of this par-
ticular construction of the coil is identical with that
of his." In this sentence it is distinctly stated that
the construction of Kennedy's induction apparatus is
identical with that of Gaulard and Gibbs'. Kennedy
accepted this statement in silence; if it had been
otherwise, he would have protested in his next appearance in print.

In order to make possible the connection of transformers in parallel, the advantages of which it may
be said Kennedy had augured, there was still much
wanting. Above all there was wanting the idea of a
transformer as meant at present, and an exact know-
ledge of its action. F. Geraldy has expressed
himself very suitably upon this point in the introduction to his report upon the trials made with the
system of Messrs. Gaulard and Gibbs.*

"La distribution de l'électricité comporte la solution d'un grand nombre de problèmes. Il ne suffit
pas de se décider en principe et lorsqu'on a choisi la
distribution en quantité (en supposant même, que
l'un des procédés puisse êıre appliqué d'une façon
exclusive, ce qui n'est pas certain), lorsqu'on a trouvé
le moyen de régler le générateur et les recepteurs
conformément au mode choisi, il reste encore à lever
quantité de difficultés, a créer et disposer beaucoup
d'organes auxiliaires." Geraldy explained distinctly
that it was not sufficient to determine only the

* La ' Lumière électrique,' vol. x. p. 496, 1883.

method of connection, but there were still a considerable number of obstacles to be surmounted before the object could be attained.

It has been a costly lesson, before the properties of transformers were known, which make them form a self-regulating system.   Even in the year 1884 do we still find Messrs. Gaulard and Gibbs on the same false track as previously.   It was in the Turin Exhibition where Messrs. Gaulard and Gibbs carried out their system upon a large scale, and where they also succeeded in gaining the interest of technical circles, and arousing general attention.

The transformers installed by Messrs. Gaulard and Gibbs in the Turin Exhibition were protected by the German patent, No. 28947, and this time again their transformers were wound with equal primary and secondary coils.   The construction of the apparatus, as already explained, made it a necessary condition that the transformers be connected in series, because only by this means could the high tension current be utilised.   It was a necessary corollary of this method of connection that the converting of the high potential of the primary circuit into low potential, was performed, not by the ratio of the number of turns in the coils of the transformers, but in a certain manner by the subdivision of the electromotive force in the circuit.

The special construction of the transformers used in the Turin Exhibition differed from the older apparatus in so far that both coils were formed of stamped out circular copper discs, which were soldered together by projecting teeth. The insulation was made of stamped-out paper discs. Both spirals were wound between one another. The building up of such coils was effected in the following manner (see Fig. 22ᴀ) :—A red copper disc was first placed on the core, then insulation, upon this a black copper disc, then again a red copper disc, and so on. Like colours of copper discs were then soldered together at the projecting teeth. In this

Fig. 22.

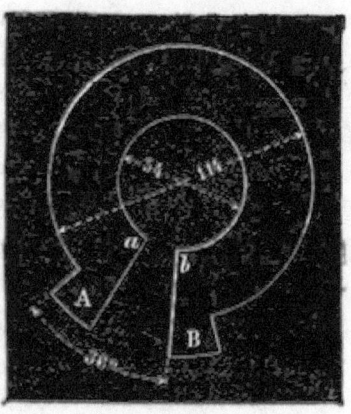

manner there were produced two spirals running parallel with one another, there only being one layer of coils. The employment of such ribbon conductors had some advantages, namely, good use of the space at disposal for coils, and rapid cooling through the projecting teeth. They had, also, disadvantages, the chief of which was, that the conductors were of bare metal, so that a fault in insulation could easily occur.

In fact, several
faults in the trans-
formers in Turin
did arise from this
cause, the action of
the coils being dis-
turbed. Further
attempts with simi-
lar coils were made,
the station houses
of Turin, Venaria,
and Lanzo being lit
for five consecutive
hours. The circuit
was about 80 kilo-
metres long, the
main lead being of
chrombronze wire
of 3·7 mm. di-
ameter. At Turin
there were 34
Edison lamps of
16 c.p. each, and
a sun arc lamp; at
Lanzo there were
nine Bernstein
lamps, 16 Swan
lamps, a sun arc
lamp, and two
Siemens' arc lamps.
In the exhibition

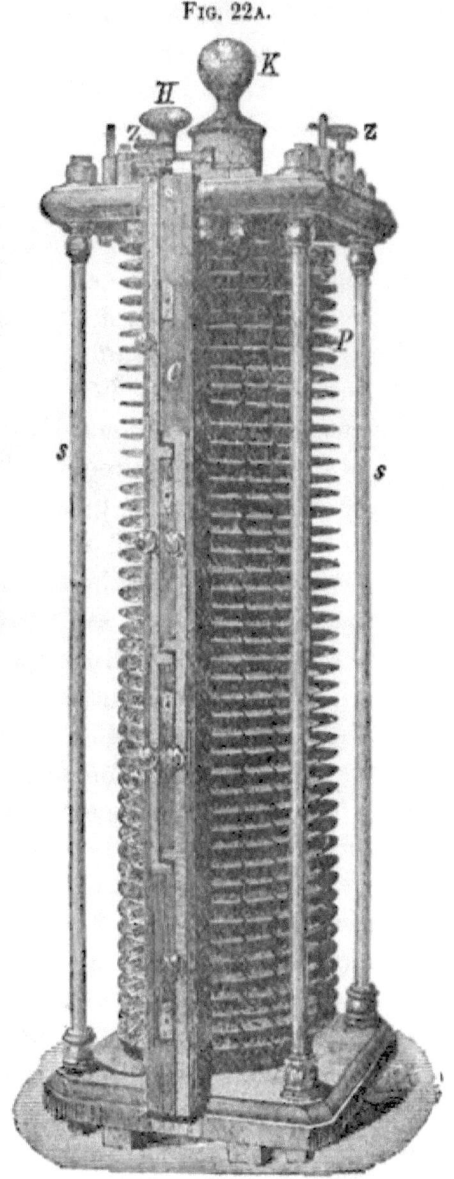

Fig. 22a.

itself, there were nine Bernstein lamps, nine Swan lamps, and a sun arc lamp. In the Figaro Kiosk nine Swan lamps were fed from a small transformer.

As already related, the trials of Messrs. Gaulard and Gibbs' system at Turin had aroused in the widest circles the liveliest interest, and, consequently, the errors of the system soon became public. Thus we find in the technical literature of that time influential voices raised against the system, and pointing out its disadvantages.

Among others, Prof. Colombo read a paper during the course of the National Exhibition at Turin, the subject being the system of Gaulard and Gibbs. While doing sufficient justice to the good points of the system, he also said that although it solved the problem of carrying the electric current to great distances, it was in no way what it was represented to be, and what it should be : a system of distribution allowing the electric current from a distant central station to be led to meet the demands of any kind of consumer without any one of these interfering with the supply of current to any other. He characterised these drawbacks sharply, and very suitably, by the remark, that in the Gaulard and Gibbs system, each consumer drew properly his supply of current from his transformer, and not from a common network of leads always self-regulating, as is the case in every large installation with continuous currents. Prof. Colombo satisfied himself with this reference to its disadvantages, mentioning also what should be striven after to make the system a perfect one, saying that the ideal electric lead system was one combining the advantages of the Edison central-station with that of Gaulard and Gibbs.

Prof. Colombo confined himself to these hints, and he must acknowledge that the means leading to the attainment of this purpose remained still to be found out.

The reproduction of this lecture by Prof. Colombo is placed before an article by Deprez in " La Lumière électrique," * in which latter the system of Gaulard and Gibbs is strongly criticised. Deprez showed that that system can have no claim to be new. He points also to the wants of the system, especially that of self-regulation, stating that the means remain still to be discovered, which would make possible the self-regulation of a system of distribution with transformers. He also says that Gaulard's system of distribution had not solved this problem, and therefore could not be held to be practically useful.

We find the same view represented in an article by H. Roux,† where he points to the enormous fluctuations which take place when the resistance in the secondary circuit is altered. Some of the figures vouching for his opinion we shall now reproduce. They were taken by M. Pietro Uzel, in Turin, in an observational way.‡

The observations are only quoted so far that the

* La ‘Lumière électrique,’ vol. xiv. p. 45.
† ‘Electricien,’ 7th March, 1885.
‡ ‘Natura,’ 25th January, 1885, p. 60.

Watts Δ I at the secondary terminals are still increasing; were they continued further the damning fact would reveal itself that as the power put in increased, the power given out would approach zero.

Taking account of these fluctuations, it is not possible to see how, as Mr. Roux says with justice, a distribution of current by this system can be made in an efficient manner. Mr. Gaulard in his reply, virtually assents to this article, but adds, that these

No.	Primary Circuit.			External Resistance of Secondary Circuit.	Secondary Circuit.			Efficiency.
	Δ	I	I Δ	W.	Δ	I	I Δ	Δ
1	23·4	12·13	283·84	1·24	15·0	12·02	180·30	63·52
2	31·4	12·13	380·88	2·00	24·0	12·00	288·00	75·62
3	53·0	12·13	642·89	3·80	45·0	11·83	532·35	82·81
4	70·0	12·13	849·10	5·50	65·0	11·75	762·45	89·80
5	93·0	12·13	1128·09	7·53	87·0	11·58	1007·46	89·31
6	107·0	12·13	1297·91	9·00	102·0	11·31	1153·62	88·88
7	126·0	12·13	1518·38	10·60	119·0	11·13	1324·77	86·66
8	145·0	12·13	1758·85	12·60	138·0	10·95	1511·10	85·35
9	159·0	12·13	1928·67	14·15	156·0	10·76	1678·66	87·03

variations could be prevented, if the cores of the transformers be shifted either by hand or automatically. Both methods would be expensive, and, besides, the automatic regulation would be unreliable.

It was at once recognised by all those interested in the subject, that this system made possible a subdivision, but by no means a distribution of current.

Before proceeding further with the history of the development of the transformer, let us for a little

while take up the question, what conditions are
necessary for a practical and rational system of
current distribution by means of transformers. As we
have already explained in another part of this paper,
the method of parallel connection, i.e., a system in
which the difference of potential is held constant, is
alone suitable. Deprez maintained in his time that the
difference of potential between the terminals of the
source of current must be kept constant. Should the
distribution be made on this principle, the resistance
of the network of leads must be very small, in order
that with full load only a very small loss of electromotive force may take place in the leads. In the
indirect system of current distribution, consequently,
the tension at the secondary terminals of the transformers must also be maintained constant.

The question is now before us, In what manner
must the primary electromotive force vary to effect
this? Consider an iron core, having on two different
parts round it, two rings of wire. This iron core may
now be magnetised by bringing near to it in the
line of its axis a permanent magnet. On drawing
the latter quickly away, an electromotive force will
be momentarily produced in both the wire rings,
and the electromotive force will be proportional to
the number of the disappearing lines of force. This
number, in consequence of the dispersion of the lines
of force, will be very different at different parts of
the magnetised core. The induced electromotive
forces in the windings of the wire will also be different. The equality of these electromotive forces,

which is so important, can only be attained if all the windings are in relatively the same position with regard to the magnetic field. The circuits of both coils being closed, the one having a current flowing through it, the other through a suitable resistance, besides the condition mentioned in the last sentence, another must be fulfilled ; this is, the internal resistance must be practically zero, i. e. the difference of potential between the terminals shall equal to all intents and purposes the total electromotive force.

We have now to examine how far the already observed constructions of transformers fulfilled these demands. A transformer in which the windings lie relatively in the same position to the magnetic field can quite well be bi-polar. All that is necessary for this is that the coils be wound on to the core next to one another; this is most simply managed in a transformer having a ratio of 1:1. This law was first determined by Maxwell. The apparatus of Strumbo shows such a method of winding already carried out.

Thus it may be seen that of bi-polar transformers, those which, with regard to the constancy of the secondary tension, are most suitable, are quite useless on account of their ratio being 1:1, although they are destined for the series method of connection.

The connection of proper transformers in parallel can only be made with such apparatus as, notwithstanding their ratio of transformation, possess windings having the same relative position to the magnetic field—this is only the case with non-polar

transformers. Besides this quality of non-polar transformers, their magnetic resistance is so low that the
condition of very low internal resistance is easily
fulfilled.

The following conditions of a self-regulating and
economical system of current distribution with
transformers result, therefore, from the foregoing
explanations :—

1. The generator of current must give a great
difference of potential as constant as possible at the
terminals of the transformers, and also independent
of the number fed.

2. The transformers must convert the current of
high electromotive force into a current of such
electromotive force as may be desired. The transformers must have a closed magnetic circuit (that is,
they must be poleless), in order that all the primary
and secondary turns shall possess, relatively to the
magnetic field, a like position, also in order that the
resistances of the primary and secondary coils shall
be so small that they cause practically no loss of
electromotive force.

Throught he fulfilment of both these conditions, it
is rendered possible to maintain the secondary tension
constant by maintaining the primary tension constant,
indifferently whether it is regulated automatically
or by hand. To suit this, the transformers must also
be arranged into distributive stations of the second
order, and derived in parallel from the main leads.

In May, 1885, a system of current distribution Zipernowsky,
Déri, Bláthy,
meeting all the just-mentioned requirements was 1885.

E

publicly brought out, giving an illustration of a
truly self-regulating system of current distribution.
This was the system of Zipernowsky, Déri, and
Bláthy.

The first two patents concerning this system date
from 18th February, 1885, and are entitled, "Improvements in the means for the regulation of
alternating electric currents," No. 34,649, by Carl
Zipernowsky and Max Déri, of Budapest; "Improvements in the distribution of alternating currents,"
No. 33,951, by Max Déri, of Vienna. The third
patent is dated 6th March, 1885, and is entitled,
"Improvements in induction apparatus for the purpose of transforming electric currents," No. 40,414,
by Carl Zipernowsky, Max Déri, and Otto Titus
Bláthy, of Budapest.

The system described in these three patents was
immediately afterwards brought forward in the
three exhibitions of Budapest, Antwerp, and London
(Inventions Exhibition), arousing in technical circles
a general and well-earned attention.

In the patent documents as well as in the earliest *
articles in the journals concerning the system, two
special forms of transformers are described, viz. that
consisting of an iron core with the wire outside, and,
secondly, that consisting of copper coils surrounded
by iron wire. The transformers shown in Figs. 24
to 28 belong to the last of these classes, that in
Fig. 23 to the first. The fundamental principle

---

* Elektricitätsverteilung aus Centralstationen, System Zipernowsky-Déri, Centralbl. f. Electrotechnik Bd. VII. S. 422.

upon which all these transformers are constructed is
that the subdivisions of the iron core run perpen-
dicularly to the copper wires. Transformers such as
are shown in Fig. 25 having a ring-shaped iron core
wound with copper wire at first employed, later

FIG. 23.

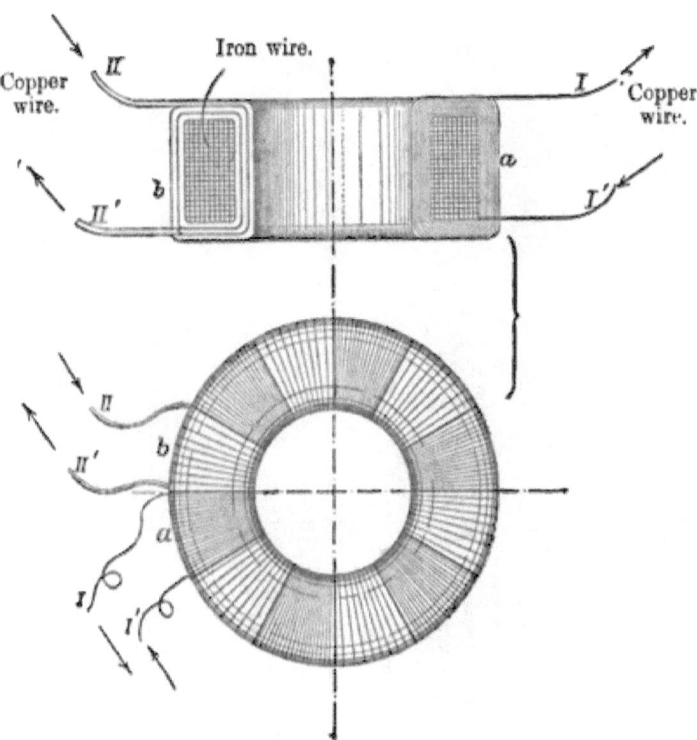

the inventors used in preference the form represented in Fig. 23.

In all these forms the principle is generally
adhered to, that the magnetic resistance and the

E 2

exciting power possess for each part of the length of the magnetic circuit the same value, and thus the formation of poles with the resulting dispersion of the lines of force is avoided.

This system procured for itself universal recognition, but especially in the Budapest Exhibition.

FIG. 24.

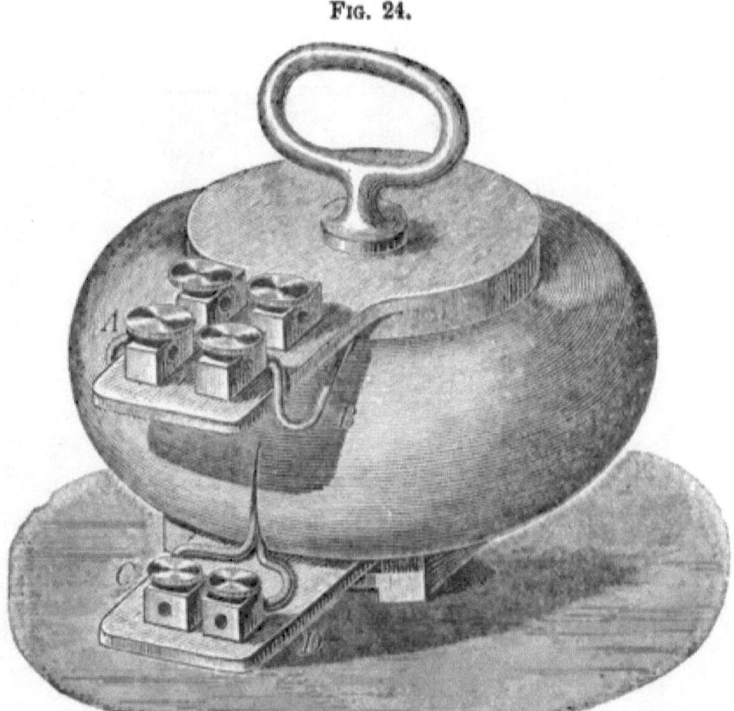

There several exhibits within a radius of 1,300 metres were lit from a common central station. The several circuits were quite independent of one another, and lamps could be extinguished or lit in

FIG. 25.

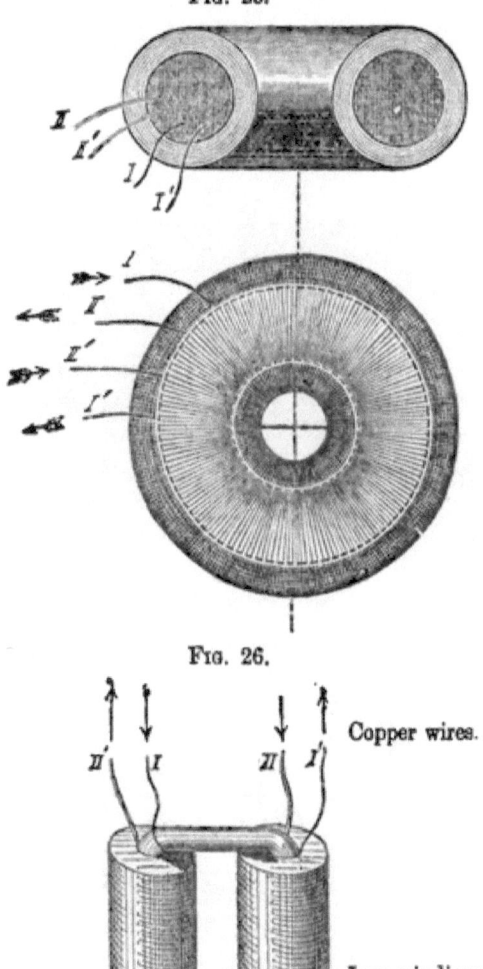

FIG. 26.

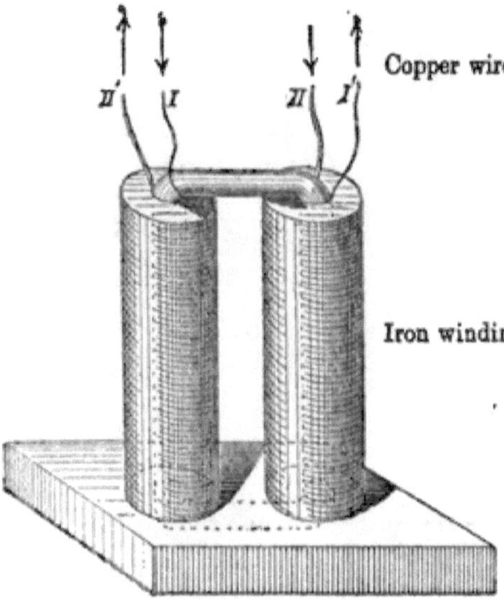

Copper wires.

Iron windings.

FIG. 27.

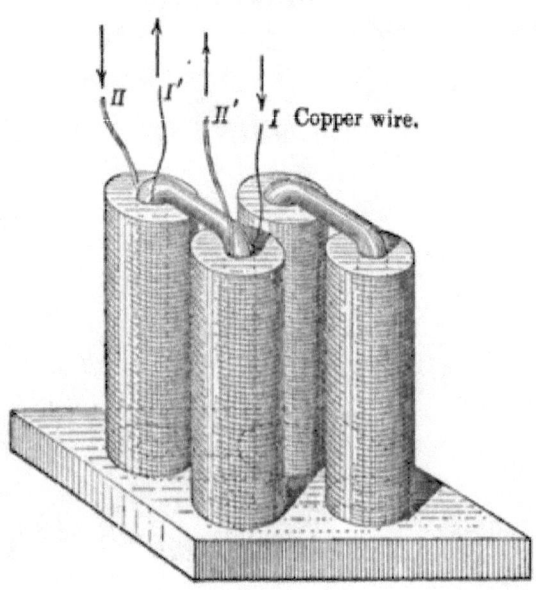

FIG. 28.

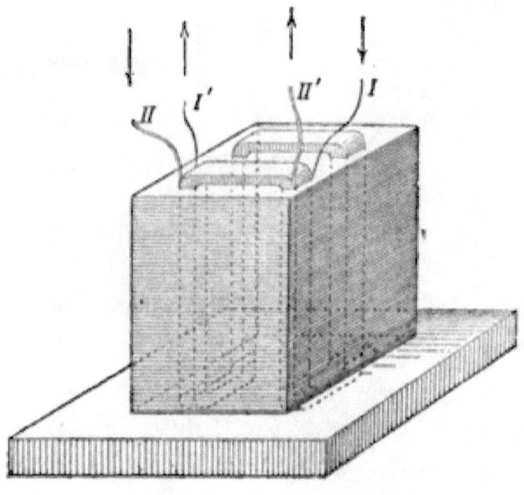

any one of them without anywhere producing a change in the intensity of the light, which could be perceived.

It was, therefore, in the year 1885 that the problem of current distribution by means of transformers was solved in a truly practical manner. The ideas which led the inventors to this thoroughly successful solution were then so unknown to practical and theoretical electricians, that it was long ere they

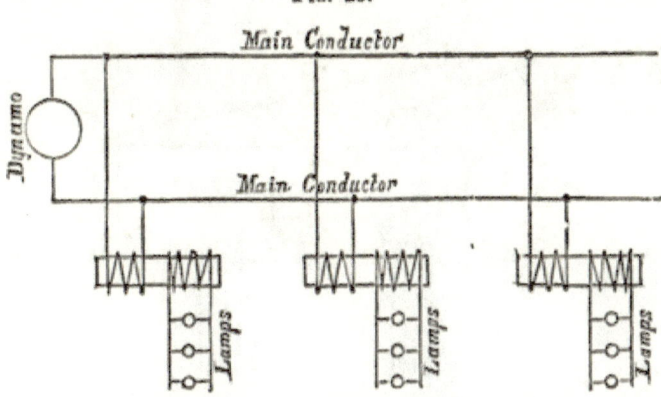

FIG. 29.

were understood and appreciated. Even in February, 1886, such an electrician as Prof. Forbes maintained in his Cantor Lectures that the parallel connection of transformers was quite impracticable. He believed, namely, that a connection such as shown in Fig. 29 was useless, because the difference of potential at the generator diminished from the machine outwards, but that a connection such as shown in Fig. 30 must be used. According to him, in a direct system of distribution each lamp should have a separate lead, and having regard to the great number of leads which would thus be necessary, he concluded that the series method of connection was the right one. One would suppose that Prof. Forbes was not aware of the weighty disadvantages of this method. However, that was not the case. He proposed, that with series connection the strength of current should be kept constant, and that each transformer should

FIG. 30.

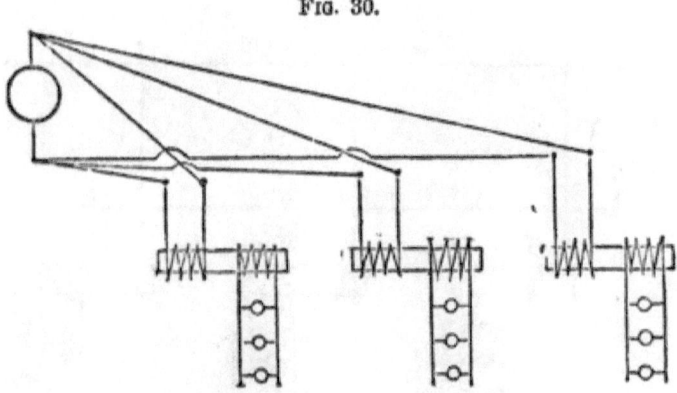

have an especial regulating apparatus—the raising or lowering of the core; which, by the way, is an arrangement impracticable in a well designed transformer. Such a regulating apparatus has lately been made automatic.

"This is," says Prof. Forbes, "the last triumph, which after a series of troublesome experiments has brought us year after year nearer to the solution of the difficulties." "I am not in a position to explain here the *modus operandi*," he says further, " but I have seen the apparatus working very satisfactorily."

This apparatus has up till now not become known. The assertion that the troublesome experiments had brought us year after year nearer to the solution of the difficulties, is quite inappropriate. Just the

FIG. 31.

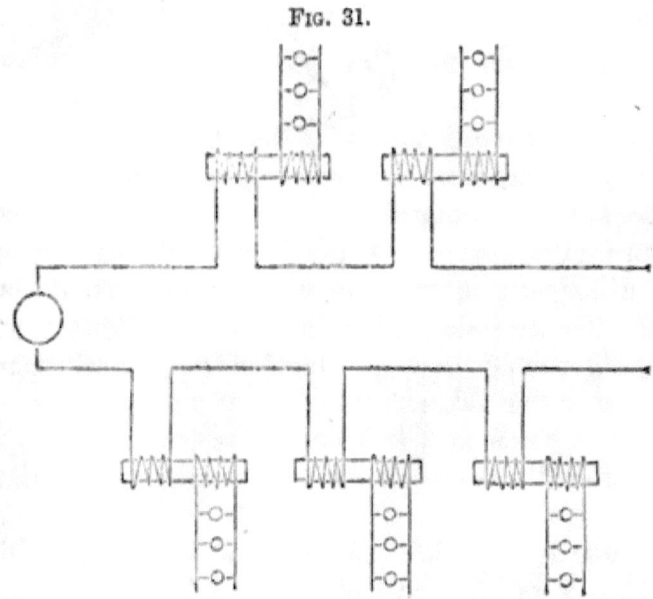

opposite is the case ; they have taken us year after year further away from the solution, until at last all was thrown overboard and a new commencement made.

Profs. Rühlmann[*] and Esson[†] also gave vent to their opinions against the connection of transformers

* 'Electrical Review,' vol. xvii. p. 157.
† 'Elektrotechnische Zeitschrift,' September, 1885.

in parallel. In a like manner Messrs. Gaulard and
Gibbs for some time after the Zipernowsky-Déri
system was known pleaded for their own method of
connection, until at last they were obliged, on account
of the unpleasant experiences at the Grosvenor
Gallery in London, to adopt the system of parallel
connection, which they then at once employed at
Tours.

There were, up till very lately, still many electricians who did not perceive the advantages of
parallel connection, just for the simple reason that
they were ignorant of the properties of the non-polar
transformer, suiting the parallel system of connection
for a rational system of distribution. Especially the
one property of transformers remained unknown to
the literature devoted to the subject up to the year
1885, namely, that in transformers properly constructed the relation between the primary electromotive force and that of the secondary, remains
unaltered notwithstanding any variations in the
current taken out; also that if the primary electromotive force be kept constant the secondary would
likewise remain constant, provided the transformer
be connected in parallel.

It had taken 30 years, until at last the way was
found leading to the desired result. We have already superabundantly explained that this direction
was essentially different from that taken by all
electricians until after Gaulard's time; that not only
the methods of connection, disposition, and regulation
of the system, but also the construction of the trans-

formers themselves had to be quite departed from, and apparatus constructed which obeyed totally other laws to those of the earlier forms.

If indeed earlier inventors proposed for other purposes magnetically-closed induction coils, the fame due to the birth of proper non-polar transformers, in which the whole of the primary and secondary turns have a like position relatively to the magnetic-field, first invented, carried out, and combined into a self-regulating systen of current distribution, belongs undoubtedly to Messrs. Zipernowsky, Déri, and Bláthy.

It would have been thought that after the direct distribution of current to glow-lamps had taken up a determined position, it would not have been difficult to discover a self-regulating system of distribution with transformers. However, the fact shows this was not the case, for after the Edison lighting system was long known, we find such electricians as Haitzema Enuma, Gaulard, and Kennedy, experimenting with the series system of connection; indeed the last of these even deters his colleagues from the attempt to run transformers in parallel, because he openly held the opinion that this method of connection was impracticable.

We have here the development of current distribution by means of transformers, as it completed itself in Europe. The American electricians however, made the matter somewhat easier. They quietly waited until the invention gave useful results in Europe, and then simply imported it.

The field to-day belongs to the parallel method of connection, and after the installation in the alkali works at Aschersleben was destroyed by flooding, there only remains a single installation with series connection, as far as we know; this is that which was fitted up in Tivoli near Rome in the year 1886. This installation however, serves only to feed an invariable number of street-lamps, and can therefore have no claim to the designation of an installation for the distribution of electric currents by means of transformers.